Cosmetics And How To Make Them

By

R. Bushby

PREFACE

THE object of this book is to provide a framework of simple theory, elementary principles, and practical methods on which the beginner in cosmetic making can build up a structure of detailed knowledge.

A large collection of formulae is now available in books and journals, but two considerations have been, to a great extent, overlooked—namely, how and why. An attempt has been made here to deal with these aspects of the subject in a simple and understandable manner.

As it is hoped that the book will be of interest to many who have no knowledge of chemistry, the small amount of chemical theory which seems indispensable to the subject has been expressed in simple language.

The apparatus described is such as will enable work on a small scale to be carried out accurately and conveniently. The formulae have been selected either as typical of a class or as good examples of a process.

R. BUSHBY.

LONDON.

CONTENTS

CHAP.		PAGE
	PREFACE	V
I.	WEIGHTS—PROPORTIONS—ELEMENTS OF CHEMISTRY	1
II.	FACE POWDER	7
III.	MEASURES—LOTIONS (INCLUDING A SKIN TONIC—AN ASTRINGENT—A MUSCLE OIL—A LIQUID POWDER) .	23
IV.	OILS AND FATS—A CLEANSING CREAM—A SKIN FOOD—A PORE CREAM .	33
V.	EMULSIONS	40
VI.	VANISHING CREAMS . . .	47
VII.	VANISHING CREAMS (CONTD.) . .	56
VIII.	POWDER CREAM—WATERPROOF CREAM—FOUNDATION CREAM . .	61
IX.	ROUGES—EYE SHADOWS . . .	68
X.	MOULDED COSMETICS . . .	72
XI.	MUCILAGES	81
XII.	NAIL POLISHING PASTES—NAIL VARNISHES AND LACQUERS . .	85
XIII.	DEPILATORIES—POWDER—CREAM—LIQUID—THALLIUM CREAM . .	91
XIV.	ALCOHOL—COSTING—VARIATIONS IN QUALITY	95
	INDEX	101

COSMETICS AND HOW TO MAKE THEM

CHAPTER I

WEIGHTS—PROPORTIONS—ELEMENTS OF CHEMISTRY

THE substances with which we have to deal may be divided into two classes, namely, *solids* and *liquids*. *Solids* are almost invariably weighed. *Liquids* may be weighed or measured.

The simplest system of weights and measures is the *metric system*. The only weights in this system of which the names need be known are—

The *gramme* (written g.);
The *kilogramme*, called kilo for short;
The *milligramme*, usually written mg.

A suitable set of metric weights usually contains the following—

ROUND WEIGHTS.
$\begin{cases} 1 \times 100 \text{ g. (not included in some sets)} \\ 1 \times 50 \text{ g.} \\ 2 \times 20 \text{ g.} \\ 1 \times 10 \text{ g.} \\ 1 \times 5 \text{ g.} \\ 2 \times 2 \text{ g.} \\ 1 \times 1 \text{ g.} \end{cases}$

FLAT WEIGHTS (under glass cover).
$$\begin{cases} 1 \times \cdot 5 \text{ g.} &= 500 \text{ mg.} \\ 2 \times \cdot 2 \text{ g.} &= 200 \text{ mg.} \\ 1 \times \cdot 1 \text{ g.} &= 100 \text{ mg.} \\ 1 \times \cdot 05 \text{ g.} &= 50 \text{ mg.} \\ 2 \times \cdot 02 \text{ g.} &= 20 \text{ mg.} \\ 1 \times \cdot 01 \text{ g.} &= 10 \text{ mg.} \\ 1 \times \cdot 005 \text{ g.} &= 5 \text{ mg.} \\ 2 \times \cdot 002 \text{ g.} &= 2 \text{ mg.} \\ 1 \times \cdot 001 \text{ g.} &= 1 \text{ mg.} \end{cases}$$

1,000 milligrammes = 1 gramme
1,000 grammes = 1 kilogramme

The small weights under the glass cover should only be handled with the forceps provided, and must be kept in their proper compartments. This enables them to be more easily recognized and provides a check, as one can see at a glance which are missing.

In order to have an intelligent understanding and grasp of any subject involving different proportions and quantities, one needs to be able to compare these proportions.

A very simple method of making comparisons, and one which is generally employed, is that of expressing the figures in terms of percentage. Percentage means "per hundred" or "in each hundred." Suppose we wish to find which of two classes of goods is the more profitable to handle, say scent and soap. On examining the books of a shop it appears that in a year the sales of scent have amounted to £400, yielding a profit of £120. The sales of soap, on the other hand, have been £600 and have shown a profit of £150. These figures do not tell us at a glance which is the more profitable article to handle, because the sales of one have been larger than the sales of the other. To compare the ratio of profit it is necessary to work out the profit on the same amount of takings.

Thus on every £100 taken for scent, £30 is profit, while on every £100 taken for soap, only £25 is profit.

We can now see which of the two articles pays the better, and by how much.

Similarly with mixtures of all kinds. In order readily to compare formulae with respect to the different proportions of ingredients, it is desirable to have them written out in such a way that the weights or volumes of the finished articles shall be the same, otherwise the issue would be confused by the fact that one formula might be written for say 1 lb., another for 30 oz., and yet another for 70 g.

Suppose the first contained $1\frac{3}{4}$ oz. of white wax, the second $3\frac{1}{4}$ oz., and the third 7·25 g. A comparison of the formulae is not possible at a glance, and certain preliminary calculations are required. In a percentage formula such calculations would be unnecessary. The formulae have only to be placed side by side for exact differences to be immediately appreciated. Another important point is that one memorizes easily and almost unconsciously the usual proportions of the various ingredients which are used.

For example, here are three formulae calling for ingredients which we will call "A," "B," and "C."

Ingredients	Formula No. 1	Formula No. 2	Formula No. 3
"A"	10·0	12·5	8·0
"B"	44·0	40·0	50·0
"C"	46·0	47·5	42·0
	100·0	100·0	100·0

One can see at a glance in what respects they differ, and, if the products are available, it is easy to link up their varying qualities with the different proportions of the various ingredients. By using the expression "per cent," one can now refer to the various proportions by one number instead of having to say "so much 'A' in so much of the product." The symbol % is employed to signify per cent. Thus formula No. 1 contains 10% of "A," formula No. 2 has 12·5% (or $12\frac{1}{2}$%) and formula No. 3 contains 8%.

ELEMENTS OF CHEMISTRY

In order to understand the nature of the materials used, and the way in which they act on one another, it is essential to have some idea of chemical constitution. All matter—solid, liquid or gas—is composed of minute particles

ELEMENTS OF CHEMISTRY 5

called "molecules," and the properties of a substance depend on the composition and constitution of the molecules of which it is built up. To enable those with no knowledge of chemistry to visualize this and understand its significance we will take an analogy.

Imagine a factory which turns out miniature toilet sets, thousands and thousands of tiny brown boxes each containing a nail file, three emery boards, and two orange-wood sticks. I visit this factory, and see from a distance huge stacks of these sets looking like solid brown masses. On a closer inspection, however, I find that these solid-looking masses are composed of little units, each exactly alike. On opening one out I find the three emery boards, the nail file, and the two orange-wood sticks. It does not need a mathematician to deduce that 100 sets will contain 100 nail files, 300 emery boards and 200 orange-wood sticks. If I now weigh the articles and find that a nail file weighs 4 g., an emery board $\frac{1}{2}$ g., and an orange-wood stick $\frac{1}{4}$ g., then it follows that each set contains—

4 g. of nail file.
1$\frac{1}{2}$ g. of emery board.
$\frac{1}{2}$ g. of orange-wood stick.

100 would contain 100 times these quantities, and any given number or weight exactly the same proportions.

Now, suppose I have before me a block of

stearic acid. It appears to be a solid white mass. It is impossible, even on a careful inspection, to see any units corresponding to the units of which the stack of toilet sets was composed. Is this because there are no such units, or because they are too small to see? The science of chemistry tells us that all matter is composed of units, just as was the stack of toilet sets, but they are infinitely smaller—far too small to be seen even with a microscope. These units are called molecules. Each molecule is built up of much smaller particles called "atoms." A molecule of stearic acid, for instance, is built up of—

> 36 atoms of hydrogen;
> 18 atoms of carbon; and
> 2 atoms of oxygen.

An atom of carbon weighs twelve times as much as an atom of hydrogen. An atom of oxygen weighs sixteen times as much as an atom of hydrogen. Knowing this, it is very easy to calculate the proportion of hydrogen, carbon and oxygen in stearic acid, just as we could calculate the relative quantities of nail files, emery boards and orange-wood sticks in a stack of toilet sets.

The chemical constitution of a substance means the number and kind of atoms of which each molecule is composed, and how they are arranged.

CHAPTER II

FACE POWDER

THE maker of cosmetics has to cater for many and varied tastes. No preparation, however good, appeals to everyone. This is just as well, as otherwise a few preparations would come to be selected by all and nothing else would stand a chance.

This being so, it is necessary to become acquainted with the distinctive properties of the various ingredients and then to use one's judgment in selecting those most desired. A face powder is required to have certain characteristics, but the blending of them is a matter of judgment. The following are the main features to be considered—

(a) *Adherence*. The powder must stay on for a reasonable time.

(b) *Opacity*. It must have a certain amount of covering power (to hide the texture of the skin), but not too much. Some people prefer considerable covering power, others like an almost transparent powder.

(c) *Slip*. The powder should go on easily and evenly.

A list of suitable ingredients follows, with suggested proportions, to give three different formulae.

Ingredients	Distinctive Characteristic	Suggested Proportions		
		No. 1	No. 2	No. 3
Calcium carbonate				10·0
Magnesium carbonate light	Lightness		5·0	
Magnesium stearate	Adherence	8·0	5·0	5·0
Osmo kaolin			35·0	40·0
Starch, rice		40·0	30·0	—
Talc	Slip	15·0	15·0	10·0
Titanium dioxide	Opacity		10·0	
Zinc oxide	Opacity	35·0		30·0
Zinc stearate	Adherence	2·0		5·0
		100·0	100·0	100·0

Each vertical column represents a formula.

The three formulae are intended as suggestions and to give an idea of the proportions which might be used. No mention is made of starches other than rice starch, as this is the best form for face powder. Other varieties may be substituted for some or all of the rice starch if cheapness is an important consideration.

Titanium dioxide has the same property as zinc oxide—opacity—but to a more marked degree, so that less is required.

Make up powders from the three suggested formulae, and carefully compare the products. Then try altering the proportions. Note particularly the effect of increasing or decreasing the proportion of talc, and also the difference caused by the increase or decrease in the proportions of zinc oxide or titanium dioxide. It will be necessary to sift these experimental

FACE POWDER

batches in order to judge their texture, etc. All quantities should be worked out for a final product of 100 g., and this is a convenient quantity to make. If, in the course of experiment, additions are made, the formula eventually arrived at should be converted, by calculation, into a percentage formula.

Colouring Materials

Colouring matters may be divided into three groups, *pigments, lakes, and dyes.*

The following will give a comprehensive range of colours—

Pigments.	Armenian bole. ⎫ Burnt sienna. ⎬ Ochre. ⎭	Earth colours consisting of iron oxide and hydrated iron oxide.
	Cadmium yellow. Ultramarine.	
Lakes.	Orange. ⎫ Red. ⎬ ⎭	Usually ordered by number from the colour manufacturers, from whom samples can be obtained.
Dyes.	Tartrazine. Oil orange.	Foodstuffs quality.

The actual composition of lakes is unfortunately not usually disclosed by the manufacturers, and some lakes are definitely poisonous. However, certain firms specialize in colours for cosmetics, and any such firm will supply suitable products. Dyes have the advantage of being, as a rule, non-poisonous, since they do not contain a metallic element. Also, much less is required to produce the same depth of colour. Lakes, however, are tending to displace dyes on account of their greater convenience. When a lake has been selected, always keep to the same firm for supplies as other makes may be different.

Apparatus

The necessary apparatus for making face powders will consist of—

    Scales and weights.
    Pestle and mortar.
    Sieve.
    Spatula.
    Boiling tube or flask (for dyes only).

For accurate work two pairs of scales are

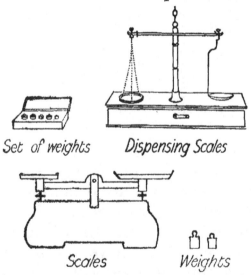

Set of weights   Dispensing Scales

Scales   Weights

necessary—one weighing up to about 50 g. of the type used for dispensing, or, better still, a chemical balance, and a larger scale weighing up to 2,000 g. Two pestles and two mortars are recommended—a glass pestle and mortar

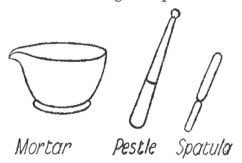

holding about 200 g. and a composition pestle and mortar holding about 3,000 g.

For sieving small quantities, silk is to be

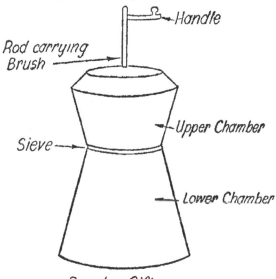

Powder Sifter

recommended. This can be held in position either by means of two wooden rings, as in the ordinary kitchen sieve, or tied over a mortar or basin. For laboratory work, sifters with a wire mesh are obtainable. These are circular and consist of two chambers, an upper and a lower. The sieve is soldered on to the bottom of the upper portion, the lid of which is pierced by a vertical rod carrying a brush, with a handle on the top for rotating the brush. There are two drawbacks to a metal sieve.

(*a*) A certain amount of metal is continually being worn off the sieve, and this very slightly affects the colour of the product.

(*b*) The sieve is unsuited to very small quantities as the brush takes up some of the powder and the last portion of powder tends to lie on the sieve rather than work through.

Silk sifting is the best, but on a small scale can only be done by rubbing the powder through with the hand or a small brush, or even a pestle.

The mesh of a sieve is expressed in terms of the number of wires or threads to the linear inch. Those used for sifting powder vary from 80 to 240. For a preliminary sifting to remove coarse particles, 80 mesh (written 80#) will be sufficient. For the final sifting, 120 or 160 mesh is recommended.

Mixing the Base

In mixing powders it is customary to start with the smallest quantity, mixing it with the

FACE POWDER

next smallest quantity, and so on, so that the quantity added each time is not greatly in excess of what is already in the mortar. The difference in weight of the various face powder ingredients is, however, largely offset by their lightness (bulk) so that, in this case, it makes practically no difference in what order the ingredients are taken.

Place the ingredients in the mortar, and stir with the pestle. Mix all together thoroughly, and then sift the product. For the first sifting a coarse sieve will do, say 80 mesh if a metal one. All the ingredients should be in the form of a fairly fine powder. If any ingredient contains a considerable proportion of grit or lumps, it must be sifted first. Otherwise so much would be removed on sifting the mixture that the final proportions would be altered. When the base has been sifted, give it a final mix, as the finer particles always tend to come through first.

Triturations

Before being used for an actual batch of face powder, the pigments and lakes must be thoroughly incorporated with some white powder. The first stage is therefore to prepare what are known as "triturations." A trituration is a dilution made by triturating or grinding together.

A convenient strength for a primary trituration

of pigments and lakes is 25%. Using the white base already made, that will be—

Pigment or lake	.	.	.	.	25
White base	.	.	.	.	75
					100

For experimental purposes, a batch of 10 g. of each will be enough. Obviously this calls for—

Pigment or lake	.	.	.	.	2·5 g.
White base	.	.	.	.	7·5 g.
					10·0 g.

METHOD. Weigh out the two ingredients. Place the pigment or lake in a mortar, and add about an equal quantity of the white base. Grind these well together until no flecks of colour or white particles are visible. Add the remainder of the white base, and repeat the process. Finally, sift through the finest silk sieve. Do this with all the pigments and lakes, put the products in boxes, tins or bottles, and label—

TRITURATION
of
....
25%.

Before actually blending the colours it will be desirable to make still more dilute triturations, except in the case of the earth pigments. A 5% trituration is suggested. This means 5 g. of the original colour in 100 g. of the product. The primary trituration contains 25 g. of colour

FACE POWDER

in 100 g., that is, 5 g. of colour in 20 g.; therefore, 20 g. of the primary (25%) trituration and 80 g. of the white base are required to produce 100 g. of a 5% trituration.

Mix these in the same way as the primary trituration. That is, add to the 25% trituration about an equal quantity or slightly more of the white base. Mix thoroughly, and add the remainder. Again mix till no specks of colour or white are visible.

Dyes

We now come to dyes. *Dyes must always be added in solution.* Some are soluble in water, some in alcohol.

Tartrazine

One of the most useful dyes for colouring face powder is tartrazine, which is soluble in water to the extent of about 1 in 20. A suitable strength of trituration for this and for other soluble dyes is 1%.

To prepare 100 g. of a 1% trituration, take

Tartrazine	1 g.
White base	99 g.
	100 g.

METHOD. Dissolve the tartrazine in about 20 ml. of warm, distilled water. (For heating small quantities of water, a boiling tube or small flask will be found very convenient. This is quick and clean, and reduces evaporation to a minimum.) A clear solution should be

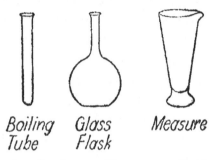

Boiling Tube   Glass Flask   Measure

easily obtained, but, if any of the particles remain undissolved, these must be filtered out.

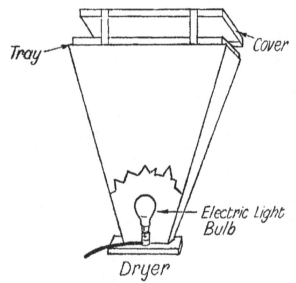

(Cut away to show source of heat)

Place about a third of the powder in the mortar. Pour on the dye solution, and stir until the powder becomes wet. Add the remainder

FACE POWDER

of the powder, and mix thoroughly. Spread the damp powder on a sheet of paper, and place in a suitably warm place to dry. An illustration of a dryer suitable for small quantities is shown on page 16. This, or something along the same lines, is easily constructed. The heat of an ordinary electric light bulb at a distance of 30-40 cm. is safe and effective.

If there is any doubt about the powder being dry, check the weight, which should be 100 g. When dry, return the powder to the mortar, reduce it to a fine powder, and finally pass it through a fairly fine sieve—120 or 160 mesh. Label this—

TRITURATION
of
TARTRAZINE
1%

Eosin is sometimes used for colouring face powder, but powders coloured with it easily bleach in the light. Also the trituration has a slight bluish cast.

Oil Soluble Orange Dye

This will give a useful pink, with only a slight suggestion of orange, and such a dye is fairly soluble in warm alcohol. Take the same proportions as for tartrazine.

METHOD. Place the powder in a mortar. Gently heat the dye in a tube or flask with about 20 g. of alcohol[1] until all or most of it appears to have dissolved. Filter directly on

[1] See page 95.

to the powder, but, as far as possible, let any undissolved portion remain in the tube. Stir the dye solution into the powder. Then add a further portion of alcohol to any residue which may remain in the tube or flask, and reheat. Pour the resultant solution over the filter paper, so that the second portion tends to dissolve also any dye remaining on the filter paper from the first portion. Now thoroughly mix the damp powder and spread it out to dry. When dry, finish in the same way as the trituration of tartrazine.

Label—
TRITURATION
of
OIL ORANGE
1%

The selection of triturations now made should be sufficient to produce any shade desired. The range of shades is so great, however, and the variation between one maker's conception of a given shade and another's so wide, that there is little point in giving formulae. The following are merely suggestions—

		No. 1	No. 2	No. 3
Trituration of tartrazine, 1%		1·0	—	—
,, Armenian bole, 25%		1·0	2·0	—
,, burnt sienna, 25%		—	2·0	—
,, yellow ochre, 25%		10·0	20·0	—
,, cadmium yellow, 5%		—	—	15.0
,, red lake, 5%		2·0	—	—
,, ultramarine, 5%		—	—	10·0
White base		86·0	76·0	75·0
		100·0	100·0	100·0

When these have been made up, try mixing various proportions according to your own idea, and try your hand at matching. Note that blue and yellow give green, and that various shades of green can be obtained by altering the proportions.

MATCHING FACE POWDERS

When attempting to match any given shade select the triturations you think will be required. Weigh out small quantities, say, 2 g. of each, and a larger quantity, say, 10 g. of white base. Place these on separate pieces of paper with the amount marked on each. Take a small quantity from the 10 g. of white base, and add the colours little by little from the quantities that have been weighed. Mix after each addition, and compare with the sample. Keep adding whatever colours appear to be required or more white base till the required shade is obtained. Then weigh carefully what is left of the triturations and the base. The difference between the weights at the commencement and what is left will obviously be the quantity used of each. The result gives the required formula.

ADDING THE PERFUME

The simplest way, when working with small quantities, is to add the perfume at the same time as the colour (triturations). When there is sufficient powder to cover well the bottom of the mortar, drop the perfume about, so as to distribute it roughly. Then triturate till it appears to have been taken up by the powder and no damp portions remain on the mortar or pestle. Here, as in practically all mixing in a mortar, the mortar and pestle must be well scraped at intervals.

On a large scale, perfumes are frequently sprayed on, and the same applies to dye solutions. This method is not readily applicable to small-scale work, and the simple method of mixing in a mortar, followed by sifting, will be found satisfactory.

When only one perfume is used, it may be better to add it when making the white base. A perfumed base, however, is not suitable for making dye triturations, since most of the perfume would be lost on drying. The advantage of keeping a perfumed base is that the perfume has a chance to mature. Actually, the most suitable ingredient to add the perfume to is light magnesium carbonate. If added to this and the product stored in the dark for some time, the perfume will improve.

The blending of perfumes is best left to the perfumer, who will supply a suitable concentrated perfume of any type required. However, if you only need something quite simple to experiment with, try an essential oil, such as oil of rose or oil of lavender, or a mixture of essential oils. About 10 drops will be sufficient for 100 g. This proportion is, of course, for the mixed base. If you are perfuming the light magnesium carbonate first, as suggested, take the corresponding amount. Thus, if the base contains 5% of light magnesium carbonate, the proportion will be 10 drops of perfume to 5 g. of light magnesium carbonate.

When both colour and perfume have been incorporated, sift through a fairly fine sieve, say 120 or 160 mesh, and give a final mix. Some sifters have a stirrer in the lower chamber, on the same vertical rod as the brush. This mixes the powder as it comes through the sieve, and obviates the necessity for a final mix.

CHEMISTRY OF THE INGREDIENTS

Some of the ingredients we have been using will be already familiar to the reader. Starch, for instance, is a household commodity. Talc is commonly known as French chalk. Zinc oxide and magnesium carbonate are in common use.

The two latter names, being chemical names, indicate the composition of the "molecules" which compose the substance. All the ingredients in our face powder base are simple substances (not mixtures), so all the molecules of which each is composed will be alike. A molecule of zinc oxide consists of one atom of zinc and one atom of oxygen. A molecule of titanium dioxide consists of one atom of titanium and two atoms of oxygen. These are both metallic oxides. A molecule of magnesium carbonate is composed of magnesium, carbon, and oxygen. Calcium carbonate contains calcium in place of magnesium. The molecules of magnesium stearate are each equal to two molecules of stearic acid in which two hydrogen atoms (one in each) have been replaced by one atom of magnesium. Zinc stearate has a similar constitution, but contains zinc in place of magnesium.

It will be remembered that the molecule of stearic acid contains a large number of atoms (compare it with zinc oxide, which has only two), and that eighteen of these are carbon atoms. Substances with a number of carbon atoms to the molecule (and some with only one or two) are known as "organic" compounds. Substances with no carbon atoms in the molecule are known as "inorganic." Carbonates are also considered inorganic.

Perhaps the main difference in this distinction from our point of view is in the keeping properties. The stearates, for instance, tend to go rancid. The oxides and carbonates cannot possibly do so. Starch is organic, but remains good as long as it is kept dry. Its molecules are composed of carbon, hydrogen, and oxygen atoms. Osmo kaolin and talc are inorganic. They are both obtained from the earth, and are very stable.

CHAPTER III

MEASURES—LOTIONS (INCLUDING A SKIN TONIC—AN ASTRINGENT—A MUSCLE OIL—A LIQUID POWDER)

LIQUIDS may be weighed or measured.

The only measures the names of which need be learnt are—

The *millilitre* or *mil* (written ml.).
The *litre*.

The millilitre is also referred to as a *cubic*

Conical Measure
with Stirring Rod

Cylindrical Measure

centimetre (written c.c.). 1,000 millilitres equal 1 litre.

Measures are either cylindrical or conical.

Cylindrical measures are more suited to very accurate work as they admit of smaller graduations. Conical measures are more convenient to use where accuracy is not so important.

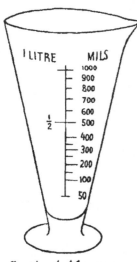

Conical Measure

RELATION BETWEEN WEIGHTS AND MEASURES

1 ml. of water weighs 1 g. A ml. of any other liquid may weigh more or less. Water is therefore a standard. It is said to have a *specific gravity* of 1. Specific gravity is usually written S.G.

1 ml. of glycerin weighs 1·26 g. In other words the S.G. of glycerin is 1·26.

1 ml. of 90% alcohol weighs 0·834 g. Therefore the S.G. of 90% alcohol is 0·834.

Formulae are frequently written with the quantities expressed only in numbers, with no indication of what these numbers refer to, thus—

    Glycerin. . . 50
    Water . . . 50

This might mean—

  Glycerin . 50 ml.   or   Glycerin . 50 g.
  Water . 50 ml.   Water . 50 g.

MEASURES, LOTIONS, ETC.

Supposing one of the ingredients is a solid, and the formula reads—

    Boric acid . . 1
    Glycerin . . . 50
    Water . . . 50

This might mean

    Boric acid . 1 g.
    Glycerin . 50 g.
    Water . 50 g.

It cannot mean

    Boric acid. 1 ml.
    Glycerin . 50 ml.
    Water . 50 ml.

since it is not practicable to measure a solid. Logically, the first is the only possibility, but actually what is probably meant is—

    Boric acid . . 1 g.
    Glycerin . . . 50 ml.
    Water . . . 50 ml.

This is not consistent, but it is nevertheless practical.

In this country it is customary to assume that liquids will be measured, and the following rule is frequently understood and sometimes misunderstood— "Solids by weight, liquids by measure." This means that when the substance is a solid the number refers to weight, and when the substance is a liquid the number refers to measure. Obviously there must be an accepted relationship between weights and measures. For this purpose it is reasonable to consider 1 g. as the equivalent of 1 ml., since 1 g. of water measures 1 ml.

LIQUID PREPARATIONS

Liquid preparations may be divided (from a physical point of view) into three main classes—

1. Solutions.
2. Suspensions.
3. Emulsions.

Emulsions will be dealt with in Chapter V.

1. Solutions

From a cosmetic point of view, these also usually fall into one of three groups—

(a) Solutions in water.
(b) Solutions in alcohol.
(c) Solutions in oil.

Since water and alcohol are miscible in all proportions, there is no sharp dividing line between groups (a) and (b).

Solution is assisted by three methods—

(i) Breaking up the particles of the substance to be dissolved, so as to present a larger surface to the action of the solvent.

(ii) Repeatedly bringing fresh portions of the solvent into contact with the substance to be dissolved.

(iii) Heating, to increase the solubility temporarily.

(i) Breaking up the particles may be done mechanically, by grinding, or by making a solution in another liquid in which the substance to be dissolved is more soluble than in the final solvent, and mixing the two liquids. The dissolved substance may be thrown out of solution in the form of very small particles when the liquids are mixed, but, owing to the increase in surface, solution in the second liquid is facilitated. It is essential that the two liquids should be miscible.

(ii) This is usually done by shaking.

(iii) Heat may be applied in the case of substances such as boric acid which are soluble only slowly in cold water but more rapidly in hot. Boric acid is soluble to the extent of 1 in 25 of cold water, but 1 in 3 of boiling water.

(a) *Solution in Water. A Skin Tonic*

Boric acid . . .	1·0 g.
Oil of neroli	0·1 ml.
Oil of rose	0·1 ml.
Distilled extract of witch hazel .	10·0 ml.
[1]Alcohol	5·0 ml.
Distilled water . . *to*	100·0 ml.

Here we have one solid and two essential oils to dissolve.

METHOD. Heat the boric acid with 20 ml. of distilled water until it has all dissolved. Add 60 ml. of distilled water and the distilled extract of witch hazel. Dissolve the two essential oils in the alcohol, and pour the solution into the other ingredients, stirring well. Then make up to 100 ml. by adding more distilled water.

[1] See page 95.

The above formula is selected as being typical of a popular type of preparation, but the proportions may be varied. Notice what happens when the solution of essential oils is added to

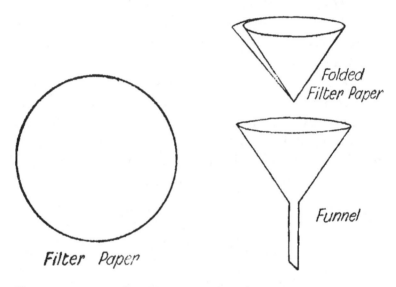

the water portion. Depending on the proportion of oils, one of three things will happen—

(1) The liquid may go very cloudy, and small globules of oil tend to come to the top.
(2) The liquid may become opalescent (cloudy), but without any definite separation of oil.
(3) The liquid may become quite clear.

This formula will probably result in condition (1), and in this case the product must be filtered. Stir in 2 g. of talc, and filter, using a funnel and filter paper. It may be necessary to pass it through two or three times.

If condition (2) is obtained, it can either be left opalescent or filtered to clear, as desired.

If an appreciable amount of oil is thrown out, reduce the proportions (for the sake of economy).

An opalescent liquid is often preferred to a clear one.

A better way of obtaining a standard opalescence, if this be desired, is by adding tincture of benzoin. This should be added to the filtered product either alone or, preferably, mixed with a little alcohol or glycerin. About 0·1% will be sufficient.

(b) *Solution in Alcohol. An Astringent*

Oil of lavender	0·4 ml.
Oil of bergamotte	0·4 ml.
Oil of neroli	0·4 ml.
Diethyl phthalate	1·0 ml.
[1]Alcohol	65·0 ml.
Distilled water	to 100·0 ml.

Here the solvent is a mixture of alcohol and water. All the ingredients are much more soluble in alcohol than in water, and in strong alcohol than in weak.

METHOD. Dissolve the essential oils and the diethyl phthalate in the alcohol, and then add the water. Filter if necessary, adding a small quantity of talc first, as in filtering the skin tonic. This absorbs any excess of essential oils.

[1] See page 95.

(c) *Solution in Oil. A Muscle Oil*

Almond oil	20·0 ml.
Nipagin m.	0·15 g.
Essential oil of camphor . .	0·5 ml.
Oil of rose	1·0 ml.
Castor oil . . . to	100·0 ml.

In this case there is a solid to be dissolved in a mixture of oils.

METHOD. Add the Nipagin m. to the almond oil and heat gently, with occasional stirring till dissolved. Allow the product to cool, and then add the essential oils and make up to the required volume with castor oil.

Nipagin m. is a preservative. It may be added in the above proportion to any preparation containing animal or vegetable oils or fats.

2. Suspensions

Watery liquids containing powder in suspension are frequently known as "liquid powders." Sunburn lotions usually belong to this class.

Since the powders are not soluble, they must be as finely divided as possible. There are three stages in preparing such a product—

(a) Mixing the powders.
(b) Sifting the powders.
(c) Incorporating the liquid.

(a) The powders for a suspension should be prepared in the same way as face powders. That is, any pigments, lakes or dyes should be incorporated by thorough trituration (grinding). In the case of dyes, should the dye be soluble in the liquid, it is better to dissolve it separately. As a rule, water-soluble dyes are not used in liquid powders, since these are preferred with only the powder coloured, and the liquid colourless. If the dye is soluble in alcohol but not in water, it should be added in solution to the powders, and the powders should be dried before sifting.

(*b*) Sifting should be done immediately before adding the liquid.

(*c*) The method of incorporating the liquid depends on the type of mixer. In a laboratory, where a pestle and mortar are used, the sifted powder should be placed in the dry mortar. On top of this should be poured the glycerin, if any, and sufficient of the aqueous liquid to produce a fairly thick cream on rubbing together. The amount of liquid added at this stage is of some importance, but must be determined experimentally. Should too little be added, a lumpy paste will result which takes a good deal of rubbing down. Should too much be added, there is too little friction to produce the smooth cream which is desired.

After the initial creamy suspension has been made, more water may be added, a little at a time, till the product is pourable. At this stage perfume and alcohol may be added, and the product made up to the required volume.

A LIQUID POWDER

Armenian bole	0·005 g.
Yellow ochre	0·02 g.
Light magnesium carbonate	1·0 g.
Zinc oxide	10·0 g.
Glycerin	5·0 ml.
Distilled water to	100·0 ml.

METHOD. Triturate the pigments thoroughly with a little of the light magnesium carbonate. Add the remainder of the light magnesium carbonate and the zinc oxide. Mix thoroughly and sift. Return to the mortar. Add the glycerin and about 5 ml. of distilled water, which may be used to rinse out the glycerin measure. This is important, as glycerin is very viscous and tends to stick to the measure. Rub down till a smooth cream is produced. Then add more water, little by little, till the product is pourable. Pour into the measure, and rinse out the mortar with a further portion of water. Add this to what is already in the measure. At this stage any desired perfume may be added, about 0·2% being usually sufficient. Now add water to the contents of the measure to bring them up to the required volume, and give a final stir with a stirring rod.

CHAPTER IV

OILS AND FATS—A CLEANSING CREAM—A SKIN FOOD—A PORE CREAM

OILS, fats, waxes, alcohol, and glycerin are all organic substances.

The word "oil" includes a great variety of substances which are so different that it is very difficult to say what particular property they have in common unless it be their "oiliness," which does not get us anywhere.

The essential oils, for instance, are quite in a class by themselves. Examples of these are rose, lavender, neroli, etc., and they form the basis of most natural perfumes. They are usually very soluble in alcohol and slightly soluble in water.

Another class of oils is that of the mineral oils, which are obtained principally from petroleum. Petroleum yields a class of substances known as "hydrocarbons." Hydrocarbons are substances whose molecules are composed entirely of carbon and hydrogen atoms. Although organic, they are very stable, which is really the meaning of the word paraffin. The term paraffin or paraffin hydrocarbons covers a range of substances, some of which are liquids and some solids (some hard solids and some soft solids). They are all built up of molecules which are composed entirely of carbon and hydrogen atoms. The difference lies in the number and arrangement. The solids have more atoms to the molecule than the liquids, and the hard solids have more than the soft.

In cosmetics three kinds of paraffin are used—

(1) Liquid paraffin;
(2) Soft paraffin;
(3) Hard paraffin.

Each of these is a mixture of paraffin hydrocarbons, but those constituting liquid paraffin have few atoms to the molecule as compared with those constituting soft paraffin, while those of which hard paraffin is composed have the largest number of all.

Oils and fats of animal and vegetable origin have quite a different composition. They are said to be saponifiable because on heating with alkalis a soap is formed (and at the same time glycerin). This forms a very important distinction from the hydrocarbon oils, which are not saponifiable.

Waxes are mixtures of substances some of which can be saponified, but which do not yield glycerin on saponification.

NON-EMULSION CREAMS

These are creams with a basis of oils and fats and which do not contain any water.

OILS AND FATS, ETC.   35

A Cleansing Cream

White wax	2.5 g.
Hard paraffin	15·0 g.
Soft paraffin	2·5 g.
Liquid paraffin	80·0 g.
Total	100·0 g.

METHOD. Weigh the basin of the water bath. Add the weight of the liquid paraffin required

Tin base of Water Bath

Basin of Water Bath

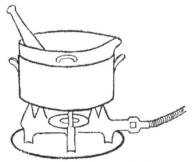

Pestle and Water Bath on gas ring

Hard wood Pestle

(80 g.). Place weights representing this total on the left pan of the scale. Pour liquid paraffin into the basin till the two sides balance. Now place the basin on its tin base containing water,

and commence to heat. While this is heating weigh the other ingredients. Add these to the liquid paraffin and stir till they are melted.

This type of cream is usually potted while warm so that it sets with a glossy surface. It should, however, be stirred from the moment it commences to set till it has begun to thicken.

A spatula is sufficient for stirring purposes for a cream of this type, and it permits of more frequent and easy scraping down of the sides than would be the case if a pestle were used. The stirring and scraping can be made practically one operation. All creams require to be frequently scraped off the sides of the basin or mortar and off the pestle, if one is being used, while setting. Otherwise lumpiness may result.

Perfume may be added on the first indication of setting. Oil of neroli and oil of lemon are popular perfumes for cleansing creams. Approximately 0·5% will be required.

A Heavy Skin Food

Wool fat	50·0 g.
Suet	5·0 g.
Almond oil	45·0 g.
Nipagin m.	0·15 g.
	100·15 g.

METHOD. Weigh the almond oil into the water bath in the same manner as in making

OILS AND FATS, ETC. 37

the cleansing cream. While this is warming weigh the other ingredients, add them to the oil, and stir till the fats have melted and the Nipagin m. has dissolved. Stir till set.

Do not heat more than is necessary. If a thermometer is available, check the temperature, which should not exceed 70° C.

It will be noted that in both examples the oils have been expressed in grammes, that is, by weight. The advantages of this are twofold—

(1) In the first place, the total amount can be readily ascertained by adding up the various quantities, whereas this could not be done accurately if the quantities referred in some cases to units of weight and in others to units of volume.

Centigrade Thermometer

(2) In the second place, when it is possible to weigh the mixing vessel the oils may be weighed into it directly. This obviates waste of time in draining measures and loss of material as well.

In cases where it is more convenient to measure, the volume is obtained by dividing the weight by the specific gravity. To convert volume to weight, multiply the volume by the specific gravity.

SUSPENSIONS IN OILS OR FATS

Solids in powder form may be incorporated in either liquid or solid combinations of oils and fats. As in the making of aqueous suspensions, the powders require to be sifted first. This can, in most cases, be done through either a silk or metal sieve. In some cases, however, the latter is inadmissible. Thus, salicylic acid and resorcin, which are used in pore creams and other medicated creams, discolour if brought into contact with a metal. In such cases only silk can be used.

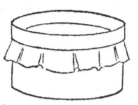

Sieve composed of two wooden rings and a piece of silk

A Pore Cream

Salicylic acid	2·5 g.
Starch (rice)	25·0 g.
Zinc oxide	35·0 g.
Castor oil	7·5 g.
Soft white paraffin	30·0 g.
	100·0 g.

METHOD. The salicylic acid must first be reduced to a fine powder by means of a pestle and mortar. Then the starch and zinc oxide are mixed with it, a little at a time being added in the usual manner. The mixed powders can now be sifted through, say, an 80 mesh silk

OILS AND FATS, ETC.

sieve. If any small crystals of salicylic acid remain in the sieve, they must be returned to the mortar and reground, and finally mixed thoroughly with the bulk.

The mortar should now be warmed and the powders placed in it. To these add the castor oil. The soft white paraffin should then be melted, and sufficient added to the powder and oil to form a smooth uniform paste on thoroughly rubbing together. As in the case of liquid powders, the object to be aimed at is to use the minimum of liquid which can be incorporated without producing a mass too stiff to be workable. In this another factor is introduced, namely, temperature. The higher the temperature the thinner will be the liquid content (oil and melted fat) and consequently the less can be used.

It is not desirable however, to raise the temperature above 40° C., as a higher temperature might lead to trouble owing to chemical changes.

CHAPTER V

EMULSIONS

Up to the present we have considered two kinds of liquid preparations, namely, those with a base of water, or water and alcohol, and those with a base of oil.

It is a matter of common experience that oil and water do not mix. By vigorous shaking or stirring a temporary mixture may be produced, but the globules of one liquid can be seen floating in the other, and these rapidly come together so that a more or less complete separation soon takes place. Even the temporary mixture bears no resemblance to the product of mixing, say, glycerin and water. In this case the product is clear, no globules of either liquid can be observed, and no separation takes place.

A very large proportion of the products with which we have to deal do, nevertheless, consist of mixtures of oils and fats with water, and these are reasonably permanent. Such preparations are not clear, but if liquid they are usually milky or creamy in appearance. These preparations are called *emulsions*, and they may be divided into two classes—

Oil in water.
Water in oil.

Shake up a small quantity of oil, such as

EMULSIONS

liquid paraffin, with a large quantity of water. Immediately after shaking, small globules of oil can be seen in the water. This temporary state illustrates an "oil in water" emulsion. The oil is broken up into separate particles which are surrounded by water. It is therefore said to be in the "disperse" phase. The water is not broken up, but forms a continuous medium surrounding the oil particles. The water is therefore said to be in the "continuous" phase.

In an emulsion the globules of oil are small (the smaller the globules the better the emulsion), and they do not come together again on standing. This is brought about by the addition of a third substance to the oil and water, called an "emulsifying agent," and also by the method of mixing.

The opposite type of emulsion—water in oil—consists of globules of water dispersed in oil.

The following puts the matter in a clearer form—

EMULSION

OIL IN WATER	WATER IN OIL
O/W ←—— abbreviations ——→ W/O	
Oil in disperse phase.	Water in disperse phase.
Water in continuous phase.	Oil in continuous phase.

The essential constituents of an emulsion are—

(a) Oil, fat, fatty acid or any mixture of these;
(b) Water or any watery liquid;
(c) Emulsifying agent.

The type of emulsion (O/W or W/O) is determined by three factors—

(1) The emulsifying agent;
(2) The proportions;
(3) The method.

Emulsifying agents can be classified according to the type of emulsion they *tend* to produce or are more satisfactory in producing. An emulsifying agent which has been selected with a view to producing one type of emulsion may, under certain conditions, produce the opposite type. It is this fact which lies at the root of many failures in making emulsions, and its possibility must always be borne in mind.

The following is a list of some of the emulsifying agents tending to produce

Oil in Water Emulsion	*Water in Oil Emulsion*
Soaps of	Soaps of
Potassium.	Calcium.
Sodium.	Zinc.
Ammonium.	
Triethanolamine.	
---	---
Tragacanth.	White wax.
Acacia.	White wax plus borax.
Mucilaginous substances.	Wool fat.
Lanette wax S.X.	Cetyl alcohol.

EMULSIONS

AN OIL IN WATER EMULSION

Triethanolamine stearate	2·0 g.
Almond oil	10·0 g.
Distilled water	88·0 ml.

This emulsion can be made by one of two methods—

METHOD 1. Heat the almond oil and triethanolamine stearate together on a water bath. (The temperature should not exceed 70 degrees centigrade.) When no solid particles are visible remove from the source of heat. Add 10 ml. of water and stir with a hard-wood pestle. The stirring should be gentle at first to avoid frothing, but more vigorous as the mixture stiffens. As it acquires the consistency of a thick cream more water should be added cautiously, and when perfectly white the balance of the water should be added.

When using this method, the changes that take place in consistency and appearance should be carefully noted. In the early stages a mixture is produced which is rather thin, and not very opaque. This is probably a coarse water in oil emulsion. A definite change takes place, however, as the mixture cools (while being stirred). The mixture thickens and whitens simultaneously, and this corresponds to the definite formation of an oil in water emulsion. If more water is added before this change takes place the chances are that only a very coarse emulsion will be produced. By coarse is meant an emulsion in which the globules are large.

METHOD 2. As an alternative method the triethanolamine stearate may be heated with 10 ml. of water, the mixture being stirred until uniform. The oil may now be incorporated, and the emulsification proceeded with as before. By this method discoloration is less likely to occur, as sometimes happens when triethanolamine stearate and a vegetable oil are heated together.

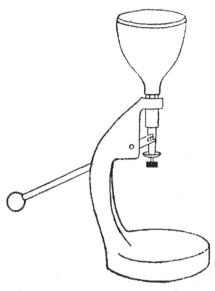

Laboratory Homogenizer used for O/W Emulsions

The product may be further improved by passing through a homogenizer if one is available. As a matter of fact, it is usually sufficient to make a rough mixture of the ingredients corresponding to the original water in oil phase, and the homogenizer will do the rest.

ANOTHER OIL IN WATER EMULSION

Lard	25·0 g.
Lanette wax S.X. . . .	7·5 g.
Distilled water . . .	67·5 ml.
	100·0 g.

EMULSIONS

METHOD. Heat the lard and lanette wax S.X. together on a water bath until both are melted. Remove from the source of heat, and stir in the water gradually. As with the triethanolamine stearate emulsion, undue stirring must be avoided until the mixture has cooled down, otherwise air will be worked in and the product will be frothy and unsatisfactory in consistency.

A WATER IN OIL EMULSION

Wool fat	12·0 g.
Soft paraffin	68·0 g.
Distilled water	20·0 ml.
	100·0 g.

METHOD. Heat the wool fat and soft paraffin together on a water bath, at a temperature not exceeding 50 degrees centigrade. When melted add the distilled water. Stir till cold. Although in this case there is no tendency to froth, there is little point in stirring much till the fats are setting. When cold, a final vigorous stir with the pestle should be given. Note the increased whiteness which this produces.

ANOTHER WATER IN OIL EMULSION

White wax	10·0 g.
Spermaceti	10·0 g.
Almond oil	59·0 g.
Borax	1·0 g.
Distilled water	20·0 ml.
	100·0 g.

METHOD. Heat together on a water bath the oil, wax and spermaceti till the solids are melted. The temperature should not exceed 65 degrees centigrade. Remove from the source of heat, and allow to cool to 60 degrees centigrade. Dissolve the borax in the distilled water by the aid of heat, raising the temperature to between 55 and 60 degrees centigrade. Pour into the contents of the water bath, stirring continuously with a wooden pestle. Stir till cold, scraping the sides of the basin and the pestle frequently with a spatula.

CHAPTER VI

VANISHING CREAMS

A VANISHING cream is an emulsion of stearic acid. The continuous phase is water with the possible addition of glycerin and alcohol. The disperse phase is stearic acid with the possible addition of oils, fats, or other fatty acids (or higher organic acids of a similar nature to fatty acids).

The emulsifying agent is a soap or a mixture of soaps of sodium, potassium, ammonium, or triethanolamine. The emulsifying agent, or part of it, is usually made at the same time as the emulsion. This is done by heating together the melted acids with an aqueous solution of sodium, potassium or ammonium hydroxides or triethanolamine.

CHEMICAL INTERACTION

Chemical constitution was defined in Chapter I as "the number and kind of atoms of which each molecule is composed, and how they are arranged."

We have now come to a case of chemical interaction. This is the rearrangement of the atoms comprising two or more molecules. Such a rearrangement sometimes occurs when the two reacting substances are merely mixed together. It usually takes place more quickly and more completely if heat is applied.

When stearic acid is heated with a solution of potassium hydroxide, each molecule of stearic acid, consisting of 18 carbon atoms, 36 hydrogen atoms and 2 oxygen atoms, comes in contact with a molecule of potassium hydroxide consisting of 1 atom of potassium, 1 atom of oxygen, and 1 atom of hydrogen. The atom of potassium changes place with a hydrogen atom in the molecule of stearic acid.

This simple rearrangement results in one molecule consisting of 18 carbon atoms, 35 hydrogen atoms, 2 oxygen atoms, and 1 potassium atom, which is called potassium stearate, and one molecule consisting of 2 atoms of hydrogen and 1 atom of oxygen, which is plain water. Expressed as a simple equation,

Stearic acid + potassium hydroxide
= potassium stearate + water.

To find the proportionate weights of the various atoms, if we call the weight of a hydrogen atom 1, a carbon atom will weigh 12, an oxygen atom 16, and an atom of potassium 39.

On this basis a molecule of stearic acid works out at 284, a molecule of potassium hydroxide at 56, and a molecule of potassium stearate at 322. These are, of course, proportionate weights. It does not matter in the least what the molecules actually weigh. It is only necessary to know the proportions.

VANISHING CREAMS

New molecules mean a new substance, and the new substances may have very different properties from the old.

The reader who is unfamiliar with chemical reactions may feel dubious about using such materials as potassium hydroxide (caustic potash), sodium hydroxide (caustic soda), and ammonium hydroxide (solution of ammonia). Were there any likelihood of any of these substances remaining in the finished product, it would be a serious matter since their action on the skin is both unpleasant and undesirable.

The chances are remote for two reasons—

(1) Stearic acid should always be considerably in excess.

(2) The temperature at which vanishing creams are made is high enough to ensure the completion of the chemical reaction.

Nevertheless, it is as well to be sure, and a very simple and interesting method is available for ascertaining when the alkali (as soluble hydroxides are called) has been completely neutralized. This consists of adding a few drops of solution of phenolphthalein to the solution of alkali. The product immediately takes on a pink tint. This pink tint persists until the alkali has been completely neutralized, when it disappears. There is no objection to using this test even when the product is intended for actual use, as the proportion of phenolphthalein is extremely small and the product is not in any way affected.

MAKING A VANISHING CREAM

In making a simple type of vanishing cream there are three factors to be considered—
(1) The emulsifying agent.
(2) The relative proportions of saponified and unsaponified stearic acid.
(3) The percentage of stearic acid in the finished product.

In order to compare the properties of the emulsifying agents, it is desirable to take equal molecular proportions and keep the other factors constant.

On referring to the figures just arrived at it will be seen that a molecule of stearic acid weighs almost exactly five times as much as a molecule of potassium hydroxide. The slight discrepancy is compensated for by the fact that potassium hydroxide always contains a trace of water.

Consequently for every gramme of potassium hydroxide we shall require 5 g. of stearic acid to convert it completely into potassium stearate. This is a soap, and provides us with the emulsifying agent.

Let us therefore start with this basis—

Stearic acid	5·0 g.
Potassium hydroxide . . .	1·0 g.

Now we require the stearic acid that is to remain as such in the final product—that is, the

VANISHING CREAMS

unsaponified portion. Let us therefore add 10 g. of stearic acid, thus—

	Saponified	Unsaponified	Total
Stearic acid	5 +	10	= 15
Potassium hydroxide	1		= 1

If we now take 84 ml. of water, this gives us a percentage formula—

Stearic acid	15·0 g.
Potassium hydroxide	1·0 g.
Distilled water	84·0 ml.
	100·0 g.

This simple example is given to show the principle involved, and how a formula is built up. Actually, it would probably yield a product rather on the watery side and with not very good keeping qualities. The first drawback can be overcome by cutting down the amount of water, the second by including a preservative. Thus—

Stearic acid	15·0 g.
Potassium hydroxide	1·0 g.
Water (distilled)	54·0 ml.
Alcohol	5·0 ml.
	75·0 g. (approx.)

Converting this to a percentage formula, we arrive at the following—

Stearic acid	20·0 g.
Potassium hydroxide	1·3 g.
Distilled water	72·0 ml.
Alcohol	6·7 ml.
	100·0 g. (approx.)

This should yield a satisfactory cream.

METHOD. Take the basin of the water bath and a wooden pestle, and weigh them together. Make a note of their weight. Weigh the stearic acid, place it in the basin, and commence to heat in the usual way. Weigh the potassium hydroxide, and dissolve it in 72 ml. of the distilled water by the aid of heat. Continue heating till the solution is almost boiling (say, 90 degrees centigrade).

Now pour it into the melted stearic acid. Remove from the source of heat, and stir vigorously till nearly cold. When fairly cool, take the basin and its contents and check their combined weight. This will be less than the total of the weights of the basin and pestle and ingredients used, owing to evaporation of some of the water. This loss must be made up by the addition of more distilled water. The simplest way is to put the required weights on the scale pan, viz. the total weight of basin, pestle, and ingredients (excluding alcohol). Place the basin and its contents on the other pan, and add water till the two balance. Stir well to incorporate the water, the temperature of which should be at least as high as that of the cream into which it is poured.

Now is the time to add the perfume, if any. Dissolve the perfume (not given in the formula) in the alcohol, and stir this in last of all, but not until the cream is almost cold. From 0·2% to 0·5% of perfume should be sufficient.

VANISHING CREAMS

Example

Net weight of basin and pestle, say .	800·0 g.
Total weight of ingredients (not including alcohol)	93·3 g.
	893·3 g.
Actual weight may be, say . . .	880·0 g.
Showing a loss by evaporation of . .	13·3 g.

Therefore 13·3 ml. of water must be added to bring the product up to the required weight.

Ammonium Cream

One molecule of ammonium hydroxide reacts with one molecule of stearic acid to form one molecule of ammonium stearate. Consequently, one molecule of ammonium hydroxide is the equivalent of one molecule of potassium hydroxide. Calculated in the same way, its weight comes to 35.

Ammonium hydroxide, however, is only obtainable in solution. The solution generally used is known as "Strong Solution of Ammonia," and it contains approximately 62·5% of ammonium hydroxide. On this basis 35 g. of ammonium hydroxide will be contained in 56 ml. of strong solution of ammonia, and each is equal to 56 g. of potassium hydroxide. We can therefore take it that 1 ml. of strong solution of ammonia is equal to 1 g. of potassium hydroxide.

Substituting strong solution of ammonia for

potassium hydroxide in the formula on page 51, we get—

Stearic acid	20·0 g.
Strong solution of ammonia	1·3 ml.
Alcohol	6·7 ml.
Distilled water	72·0 ml.
	100·0 g. (approx.)

The procedure will be slightly different owing to the fact that the alkali is volatile and in solution.

METHOD. Having noted the weight of the basin and pestle, melt the stearic acid on the water bath. Take about 65 ml. of the distilled water and heat till it is almost boiling. Pour this into the melted stearic acid. Dilute the strong solution of ammonia with the remainder of the distilled water, and immediately pour into the stearic acid and water, stirring at the same time.

The remainder of the procedure is the same as with the potassium cream.

Sodium Cream

A sodium cream can be made in precisely the same way as potassium cream. A molecule of sodium hydroxide weighs 40, as compared with 56 for potassium hydroxide. Therefore, less (by weight) will be required. In place of 1·3 g. of potassium hydroxide we shall require

$$1\cdot3 \text{ g.} \times \frac{40}{56} = 0\cdot93 \text{ g.}$$

VANISHING CREAMS

Triethanolamine Cream

As a fairly pure form of triethanolamine stearate is obtainable, there is no reason why this should not be used instead of triethanolamine itself. Either can be used.

Compared with potassium hydroxide (56), a molecule of triethanolamine weighs 149, and a molecule of triethanolamine stearate 433.

Again, substituting in the amended potassium formula, these figures would give—

(a)
Triethanolamine	3·5 g.
Stearic acid	20·0 g.
Alcohol	6·7 ml.
Distilled water	69·8 ml.
	100·0 g. (approx.)

(b)
Triethanolamine stearate	10·0 g.
Stearic acid	13·5 g.
Alcohol	6·7 ml.
Distilled water	69·8 ml.
	100·0 g. (approx.)

METHOD (a). Heat the distilled water and dissolve the triethanolamine in it. Pour into the melted stearic acid, and proceed as before.

METHOD (b). Melt the triethanolamine stearate with the stearic acid. Boil the water, and pour in, stirring till cold. Finish in the usual way.

CHAPTER VII

VANISHING CREAMS (CONTD.)

ONE should now be in a position to experiment intelligently with vanishing creams. We started with a basis of 10% unsaponified stearic acid, which was subsequently increased to 13·3%; we can now try out other percentages by adjusting the weight of the final product. Having arrived at a method of calculating the amount of alkali required to neutralize a given amount of stearic acid, the actual amount of emulsifying agent can be varied at will, although it is not usually added as a single substance. Thirdly, different emulsifying agents can be used without varying the percentage of saponified stearic acid.

The chemical explanations and resultant calculations given in the last chapter may have seemed unnecessary and possibly wearisome to those who feel that their interest lies in the practical rather than in the theoretical side. Without them, however, it would hardly be possible to experiment with any intelligence in modifying existing formulae or trying out new ones. In any case, it is far more satisfactory to understand what is taking place and why, than blindly to make up some stock formula in a mechanical manner.

VANISHING CREAMS

In comparing two products in which the percentage of free stearic acid, the nature of the emulsifying agent, and the percentage of saponified stearic acid were all different, it would be impossible to decide which of the three factors was responsible for any difference between the two products. It is only by keeping two factors constant that one can judge the effect of varying the third.

MODIFICATION OF BASIC FORMULA

To commence with the continuous phase, the two liquids frequently added are glycerin and alcohol. The former may be included up to 50%, and the latter up to 20%. High percentages of glycerin are characteristic of French creams, but are not recommended.

Glycerin is usually added with the water when making the cream, although it can be added after the cream has cooled down. Try substituting glycerin for some of the water in any of the foregoing formulae. If the percentage is very high, it is essential to add the glycerin with the water, 90 degrees centigrade being a suitable temperature for the mixture.

Alcohol must not be added until the cream is cool. Before adding it, adjust the weight and see that the cream is free from lumps, as up to the addition of the alcohol it may be reheated if necessary, but not afterwards, as the alcohol would evaporate. Perfume should be dissolved in the alcohol, and it is desirable always to include sufficient alcohol for this purpose.

Additions to the continuous phase might consist of liquid paraffin, almond oil, or oleic acid. The following is an example—

Stearic acid	15·0 g.
Almond oil	3·0 g.
Liquid paraffin	5·0 g.
Strong solution of ammonia	1·5 ml.
Glycerin	5·0 g.
Distilled water	to 100·0 g.

METHOD. Melt the stearic acid with the oils. Heat the glycerin with most of the water, and stir in. Mix the strong solution of ammonia with a small quantity of warm water and add, stirring continuously till cool. Make up to weight.

ADDING THE PERFUME

The percentage of concentrated perfume in a cream should not usually exceed 1%, and 0·5% will frequently be ample. Some perfume constituents are irritating to the skin if more than a small percentage is present. Such substances are aldehydes and ketones.

From the physical point of view concentrated perfumes may be regarded as oils soluble in alcohol. The more expensive products contain a large percentage of essential oils, while the cheaper ones are composed largely of synthetic chemicals. The latter are, however, of a similar oily nature. This fact should be borne in mind when adding the perfume to the preparation.

VANISHING CREAMS

Oily compounds or W/O emulsions will take up the perfume immediately. In the case of O/W emulsions, it may be necessary to add the perfume to the oil phase before emulsifying. If alcohol is present, the perfume may be dissolved in it.

LIQUID VANISHING CREAMS

These are made on the same principle, the water content being increased till a liquid is produced. This should not be too thin, or separation is likely to occur. Triethanolamine stearate provides an excellent emulsifying agent for this type of preparation, and is to be preferred to the other stearates for the purpose.

Take the formula given in Chapter VI, page 55, as the basis. When the cream has cooled down somewhat, add more distilled water gradually and with constant stirring till the product is just pourable.

ADDITION OF ALCOHOL TO EMULSIONS

When alcohol is added to liquid emulsions, it must be done cautiously; 20% is probably the maximum that can be incorporated without the emulsion "splitting" (sometimes called "cracking" or "breaking"). It is quite possible, however, to produce local splitting without the whole going, and this may be overlooked. Obviously when a small quantity of one liquid is poured into a larger quantity of another, the former does not immediately diffuse through the whole amount of the latter. Patches may occur in which the added liquid is present for a short time in a high concentration. This can easily result in splitting of a small patch of the emulsion, and consequent separation of some of the oil or fatty acid, which will be stirred into the bulk and not noticed till the emulsion has stood for a short time.

Glycerin can be added in almost any proportion. A small amount will be an advantage if no oils or fats are present, as it gives "softness" to the product.

As in the stiffer creams, oils may be added to the disperse phase. Such products are improved by passing through a homogenizer.

CHAPTER VIII

POWDER CREAM—WATERPROOF CREAM—FOUNDATION CREAM

UNDER the above and other names a variety of preparations are sold which are both emulsions and suspensions. That is to say, they consist essentially of a powder suspended in an emulsion. The emulsions may be of either type, so we can divide them primarily into—

(1) Suspensions in an O/W emulsion;
(2) Suspensions in a W/O emulsion.

For the purpose of classification we have included fatty acids with oils under the heading of oil in water emulsions.

It may be as well to divide (1) above into

(A) Those based on emulsions of fatty acids;
(B) Those based on emulsions of true oils.

GROUP (A)

These are powder creams with a vanishing cream base. Combinations of vanishing creams and powder may easily "roll" badly and lack smoothness, but a high percentage of glycerin tends to overcome this. Powder should not be incorporated with a vanishing cream while in the dry state, but should be rubbed down with some liquid first till a smooth cream is formed. It may, therefore, be desirable to add the whole or part of the glycerin not with the water portion when making the vanishing cream, but afterwards with the powder. If the vanishing cream portion is not cold when the powder suspension is added, this should be brought to the same temperature.

What qualities should we look for in the powder? Our base already contains a solid in the form of stearic acid. Any considerable addition of powder will cause this to "roll" when an attempt is made to work it into the skin. Consequently bulk is not desirable.

A vanishing cream itself should produce a matt appearance and also a very slight amount of tackiness so that powder tends to adhere. Hence, an adherent powder is not called for.

What is primarily wanted is "cover" or opacity. Two substances possess this property in a marked degree—zinc oxide and titanium dioxide.

Of these, titanium dioxide has about five times the opacity of zinc oxide. It also has the very important advantage of being chemically inert. Zinc oxide is always liable to react with a fatty acid to form a zinc soap—in this case zinc stearate.

If the zinc oxide is incorporated in the cold, the reaction will probably not take place for some time. There is, however, the possibility, if not the probability, that the changes would take place eventually, and would result not merely in the change of some of the stearic acid and the zinc oxide into zinc stearate, with its somewhat different texture, but also in the splitting of the emulsion and consequent separation of the water content from the solid portion of the cream. It will be remembered that soaps of zinc tend to produce W/O emulsions which are the opposite type to the one we are dealing with.

Titanium dioxide, therefore, has the two advantages of greater opacity and chemical stability, and is to be preferred.

Another property that may be found desirable is "slip," so a small addition of talc is a possible advantage.

The preliminary treatment of the powder will be the same as for liquid powders. Water soluble dyes should be avoided. The powders, when coloured, must be finally sifted and then rubbed down with glycerin. When a smooth cream has been formed, a base may be incorporated.

For a base, take ammonium stearate as the emulsifying agent. The formula in Chapter VI, page 54, gives

Stearic acid	20·0%
Strong solution of ammonia .	1·3%
To this add Glycerin	40·0%
Leaving for Distilled water . . .	38·7%
	100·0

This might be further improved by increasing the percentage of emulsifying agent and still further cutting down the water (by increasing the percentage of all the other ingredients), thus—

Stearic acid	25·0%
Strong solution of ammonia	3·5%
Glycerin	50·0%
Distilled water	21·5%
	100·0

It will be necessary to take considerably more water to commence with than is shown in the formula, say 40 ml., as otherwise it would be very difficult to work up. The difference, viz. 18·5 ml., will almost certainly be lost by evaporation.

A suggested formula is—

Talc	2·0 g.
Titanium dioxide	8·0 g.
Glycerin	3·0 g.
Distilled water	3·0 ml.
Alcohol	4·0 ml.
Vanishing cream base	80·0 g.
	100·0 g. (approx.)

METHOD. Incorporate any necessary colour with the powders, mix, and sift. Rub down with the glycerin, distilled water and alcohol. Add the vanishing cream base, and stir till uniform.

Note. In formulae such as the above, colouring agents are not given. These are left to the

POWDER CREAM, ETC.

reader's discretion, as sufficient information on the subject was given in Chapter II. The percentage required being very small, the proportions of the other ingredients will not be appreciably affected by their addition.

Similarly, perfumes may be added as desired in the manner previously suggested.

GROUP (B)

In this type the oil is usually liquid paraffin. This is preferable for the reason given for titanium dioxide, viz. chemical stability. Zinc oxide and talc are frequently used for the powder portion, and tragacanth for the emulsifying agent.

Although tragacanth is not suitable as the principal emulsifying agent in liquid preparations, it answers fairly well in stiff creams.

The zinc oxide might be replaced by a mixture of osmo kaolin and titanium dioxide. This would permit of the use of triethanolamine stearate as the primary emulsifying agent, since the possible formation of zinc stearate would be eliminated. Strangely enough, a small proportion of zinc stearate is sometimes added to this type of preparation. This increases the adherence and "waterproof" character. The addition of a definite percentage at the time of mixing, however, is quite a different matter from the gradual formation of an indefinite amount over an indefinite period.

The following is an example—

Zinc oxide	25·0 g.
Talc	10·0 g.
Tragacanth	0·2 g.
Liquid paraffin	7·0 ml.
Distilled water	45·0 ml.
Glycerin	12·8 ml.
	100·0 g. (approx.)

METHOD. Mix the zinc oxide and talc with any required colouring matter, then sift. Rub the powder down with about 10 ml. of glycerin and about 30 ml. of the distilled water. Mix the tragacanth with the remainder of the glycerin, and then add to it the balance of the distilled water to form a mucilage (see Chapter XI). To this add the liquid paraffin in small quantities, stirring continuously and emulsifying each portion before adding the next. Finally, add the emulsion to the powder suspension.

In combining emulsions with suspensions, one must, as previously stated, always bear in mind the possibility of "splitting" the emulsion, and it is sometimes necessary to experiment in order to find a satisfactory method of blending an emulsion with a suspension. In the example given the emulsion may be incorporated by rubbing (with a pestle), but in some instances the friction of the pestle and the suspended powders would split the emulsion. Owing to the thickness of the product this might be overlooked. No obvious separation would occur, but, in use, the difference between a preparation containing a thoroughly emulsified oil and one containing an oil not so emulsified would be noticeable to the disadvantage of the latter. The splitting is usually associated with a slight thinning, and this should be looked for. When this is liable to occur it may be found that beating or stirring lightly without rubbing will produce a satisfactory mixture.

SUSPENSION IN A W/O EMULSION

The selection of good emulsifying agents of the W/O type is not great, though certain proprietary ones are now available.

The following formula utilizes a zinc soap, and represents the definitely greasy type—

Zinc oxide	20·0 g.
Zinc oleo-stearate	10·0 g.
Liquid paraffin	60·0 g.
Distilled water	10·0 ml.
	100·0 g.

METHOD. Heat the zinc oleo-stearate with the liquid paraffin till it has melted. Sift the zinc oxide (coloured if necessary), place in a mortar, and pour on to it a portion of the oily liquid. Rub down till smooth, and add the remainder. Finally, incorporate the water, adding it gradually.

CHAPTER IX

ROUGES—EYE SHADOWS

ROUGES

COLOUR for the cheeks may be incorporated in a great variety of different bases. The following are the three principal types of bases—

(1) Powder.
(2) Grease.
(3) Vanishing cream.

Powder

The technique for preparing a powder rouge is not essentially different from that for preparing a face powder. Naturally, the percentage of colouring agents is much higher.

Grease

A grease rouge may be soft or hard. The simplest form of a soft grease rouge consists of a base of soft white paraffin with which is incorporated a dye or lake, usually the latter. The following may be taken as a basic formula—

Lake	15·0 g.
Soft paraffin, white	85·0 g.
	100·0 g.

METHOD. Sift the lake. Rub it down well with a small quantity of the soft paraffin in a warm mortar, till perfectly smooth. Incorporate the remainder of the soft paraffin.

Hard Grease Rouge

This may be made by substituting white wax for some of the soft paraffin in the previous formula. The addition of a small proportion of talc (and also liquid paraffin) may be considered an improvement. Thus—

Lake	15·0 g.
Talc	15·0 g.
White wax	5·0 g.
Liquid paraffin	10·0 g.
Soft paraffin, white	55·0 g.
	100·0 g.

METHOD. Mix the lake and talc, and sift. Rub this mixture down with the liquid paraffin and a portion of the soft paraffin. Melt the white wax with the remainder of the soft paraffin. Add the colour portion, and stir all together. By mixing in the vessel in which the wax has been melted the whole can be melted together, probably without the application of further heat. Stir till almost cold, and pour into containers.

Vanishing Cream Base

The base given for a powder cream should be found satisfactory. Thus—

Stearic acid	25·0 g.
Strong solution of ammonia	3·5 ml.
Glycerin	50·0 g.
Distilled water	21·5 ml.
	100·0 g. (approx.)

Alcohol soluble dyes are suited to this type of base. Lakes can be incorporated in small amounts, but do not, as a rule, produce such a satisfactory product.

Try 1% solutions of *dyes*, or stronger if desired and the solubility of the dyes allows it. Eosin can be taken as a good example. Dissolve 1 g. of this in sufficient alcohol to produce 100 ml., and filter. Label thus—

SOLUTION OF EOSIN
1% in Alcohol.

Having prepared the base, take

Vanishing cream base	80·0 g.
Dye solution	20·0 ml.
	100·0 g. (approx.)

Perfume can be dissolved in the dye solution.

METHOD. Using a pestle and mortar, add the solution to the base in small portions, stirring after each addition. Finally rub together till smooth.

In the case of *lakes*, these may be incorporated up to 2%, as in this example—

Lake	2·0 g.
Glycerin	5·0 ml.
Alcohol	13·0 ml.
Base	80·0 g.
	100·0 g. (approx.)

METHOD. Sift the lake, and rub it down with the glycerin. Add the alcohol and incorporate the base.

EYE SHADOWS

Eye shadows may be prepared in the same way as rouges, but it is not desirable to use soluble dyes. They are usually no more than suspensions of lakes or pigments and other powders in soft paraffin.

The following are suggestions for obtaining the more popular shades—

Brown . . Burnt umber
Blue . . Ultramarine with a trace of lamp black
Green . . Ultramarine and cadmium yellow

The incorporation of a white powder, such as a face powder base, will improve the consistency.

The procedure will obviously be—

(a) Mix the powders;
(b) Sift the powders;
(c) Incorporate with the least quantity of soft paraffin that will give a smooth product;
(d) Incorporate the remainder of the soft paraffin.

Heat may be employed, if desired, to facilitate mixing in stage (c). The quantity of soft paraffin required at this stage will depend on the method of mixing.

For *gold* eye shadow use gold bronze powder . 40%
For *silver* eye shadow use aluminium powder No. 334 . . . . 20%

If no vegetable or animal oils or fats are used, no preservatives will be required.

CHAPTER X

MOULDED COSMETICS

FROM the point of view of use moulded cosmetics may be divided into two classes—

(1) Those to be applied while still solid.
(2) Those which have to be melted before application.

Class (1) includes *lipsticks* and *eyebrow pencils*. Class (2) includes *depilatory wax* and *wax masks*.

From the point of view of composition, the first class may be divided into—

(a) Those containing the colouring matter in solution.
(b) Those containing the colouring matter in suspension.

There is, of course, no reason why a preparation should not come under both headings, and probably most modern lipsticks do.

LIPSTICKS

A lipstick base usually consists of a mixture of fats and waxes, with oils and stearic acid as

MOULDED COSMETICS

possible additions. A simple base could be made as follows—

Castor oil	70·0 g.
White wax	30·0 g.
	100·0 g.

This may be modified by the addition of almond oil, suet, wool fat, spermaceti, stearic acid, etc. The reader should experiment with various combinations of these and other oils, fats and waxes, but for trying out colouring agents the above simple formula will suffice. One of the most useful dyes for lipsticks is *eosin*. This can be obtained in a variety of shades and in different chemical forms. It has high tinting qualities, and is soluble in alcohol and water, though not so readily in the latter.

Ordinary eosin is not directly soluble in oil. A special form known as "acid eosin" or "bromo acid" is soluble in oils and fats, but not so readily as could be desired.

A simple method of producing what is practically an oil solution of eosin is to dissolve it first in alcohol, mix the solution with castor oil, and evaporate off the alcohol. The great advantage of an alcoholic solution is that it can be filtered, thus ensuring freedom from undissolved particles of dye stuff.

With small additions of alcohol to a melted lipstick, the evaporation is frequently left to chance, the assumption being that most of it will evaporate. If any considerable amount is added, however, the weight must be checked and heat maintained until all the alcohol has evaporated, or only a small and fairly definite percentage is left.

The basic formula can now be extended as follows—

> Castor oil . . . . 70·0 g.
> White wax . . . . 30·0 g.
> Solution of eosin in alcohol . 200·0 ml.

This last figure will depend on the strength of the solution used, and the amount of tinting property desired. Much less may be found sufficient.

METHOD. Weigh the basin of the water bath. Pour in the castor oil and solution of eosin. Mix, and commence to heat. Add the white wax. When most of the alcohol appears to have evaporated, check the weight. Heat for a little longer, and weigh again. If the weight has not altered appreciably, remove from the source of heat. If there has been a definite loss in weight, heat further till two successive weighings show no appreciable difference. Note the final weight of the product. One final stir before pouring into moulds will now be, sufficient.

Lipsticks containing eosin in solution have the interesting property of imparting a different colour to the lips from that of the stick itself. The same thing, of course, applies in the case

MOULDED COSMETICS

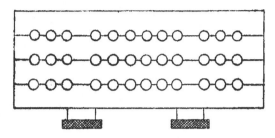

Shut Mould
View from top

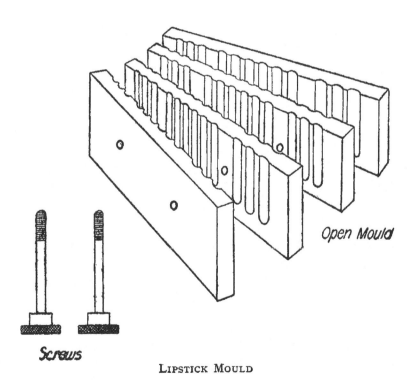

Open Mould

Screws

LIPSTICK MOULD

of rouges. Other soluble dyes may be used, but their properties of colouring the lips are hardly comparable with those of eosin.

Lakes, and frequently dyes as well, are suspended in the base. The greatest care must be taken to ensure that the colouring matter is in a very fine state of sub-division and that the product is smooth. To this end it is desirable to sift the lakes and dyes first. They must be ground down with the liquid (or most readily liquefiable) portion of the base. In our simple formula the castor oil forms an ideal medium. Place the sifted lake in a mortar, and pour on it a small quantity of the oil. Grind the two together, adding more oil as necessary. When a perfectly smooth product results, the remainder of the oil may be added. Melt the wax on a water bath, and add the oil suspension. Stir till on the verge of setting, and then pour into moulds. The mould should previously have been lubricated with liquid paraffin or almond oil. The following points should be borne in mind—

(a) *The Mould.* This requires efficient but not excessive lubrication. Use plenty of lubricant to make sure that the surface is thoroughly covered, and wipe off any surplus.

(b) *The Mortar.* To save using two utensils where one might do, and to avoid possible wastage, there is a temptation to utilize the water bath for rubbing down the powders as

MOULDED COSMETICS

well as for melting the fats. It is impossible, however, to get the same grinding action in a porcelain basin as in a mortar.

(c) *The Melted Mass.* If the mass, containing undissolved solids in suspension, be poured into the mould while too hot, there will be a tendency for the solids to gravitate towards the bottom before the mass has had time to set. Consequently, there will be an excess of colouring matter in the top of the stick. On the other hand, if the pouring is left too long the mass will set while pouring.

A suitable percentage for red lake is 25. The following is a suggested formula—

Red lake	25·0 g.
Castor oil	25·0 g.
Almond oil	20·0 g.
Wool fat	10·0 g.
White wax	20·0 g.
	100·0 g.

Some lipsticks are made by "massing" in a similar manner to that described under the heading "Mascara" (see page 79). The resultant mass requires to be moulded under pressure, but for experimental purposes this can be done by hand.

LINERS

Eyebrow pencils and liners can be made in a similar way to lipsticks. For the former, lamp black and brown umber will take the place of lakes and dyes.

DEPILATORY WAX

This product is frequently supplied in small saucepans ready for use. Consequently there is no question of turning it out of a mould and no necessity to lubricate in such cases. When it is required in blocks, slabs or cakes, however, the mould, which may be any sort of shallow tin, should be lubricated with castor oil.

Again, precaution must be taken against pouring it in too hot, not on account of the possible separation of the suspended matter, but because the basic ingredients, resins and waxes, tend to separate on cooling. Furthermore, a wax poured into the mould too hot is difficult to remove.

The following is a formula for depilatory wax—

Colophony	80·0 g.
Beeswax	20·0 g.
	100·0 g.

METHOD. This is made by merely melting together the ingredients, stirring occasionally. When the product is thick, but still pourable, give it a final stir, then allow to stand for a few seconds so that any bits of solid impurities may fall to the bottom. Now pour into moulds or saucepans, according to requirements.

WAX MASKS

These are usually moulded into small tablets, each sufficient for one application. The following is typical—

Olive oil	5·0 g.
Soft paraffin	25·0 g.
Hard paraffin	70·0 g.
	100·0 g.

METHOD. Melt the two paraffins on a water bath with the oil. When almost cool, pour into suitable moulds. These tablets are usually delicately tinted. Any oil soluble dye can be used to obtain this effect.

WATER COSMETICS—MASCARA

Water cosmetics may require an entirely different technique for their production. The following formula represents a "mass" which does not at any time assume a liquid form.

Lamp black (greaseless) . .	30·0 g.
Powdered soap . . .	30·0 g.
Powdered tragacanth . .	6·0 g.
Triethanolamine stearate . .	34·0 g.
	100·0 g.

METHOD. First mix and sift the powders. Then add the triethanolamine stearate and work in, using considerable pressure. A pestle and mortar much larger than would be required for a similar weight of the usual type of preparation should be used. After a rough preliminary mix, designed to break up the stearate and distribute it among the powders, the pestle should be grasped in such a way that its head is up against the ball of the hand. The pestle is then pressed hard against a portion of the mixture and rotated outwards (clockwise) at the same time. This method of manipulation, known as "massing," produces eventually a smooth uniform product which can be rolled out and cut into suitable shapes. The massing requires some energy (if done by hand), but must be continued until the mass begins to come clean from the mortar.

Coloured products may be similarly made by substituting pigments or lakes for the lamp black.

CHAPTER XI

MUCILAGES

A VARIETY of substances occur in the vegetable and animal kingdoms which will either dissolve or swell up in water to produce jellies, or thick, viscous liquids, known as mucilages. The products are sometimes used by themselves and sometimes in combination with O/W emulsions or aqueous (watery) suspensions.

As an example of a substance of animal origin we have *gelatin*. This is not soluble in cold water, but it is soluble in hot water. It swells up, however, in cold water, and then dissolves readily on heating. To make jelly from it, cut up the gelatin into small pieces, allow to soak in cold water for some time, then heat on a water bath. On cooling, a jelly is formed.

Many substances of vegetable origin are available. One (which is largely used) is *tragacanth*. This is a gum which occurs in flakes, but it is frequently bought in powdered form for convenience. It is only partly soluble in water, but swells to a gelatinous mass. If the flakes are used, they require considerable soaking, and finally heating, before a uniform, gelatinous mass is formed. The powder is much more convenient and handy to use, but preparations made with it do not acquire their maximum viscosity for some little time. On the addition of water to powdered tragacanth, thickening takes place very quickly, and lumps may easily be formed which are very difficult to disperse. It is desirable, therefore, to mix the powdered tragacanth first with some other liquid.

Thus, in making a simple mucilage, the powder may be diffused in a small quantity of alcohol. It is a good plan to put the alcohol first into the mortar, bottle or other receptacle, and shake or stir until the sides have been completely wet by the alcohol. This takes off any trace of moisture (water) which may cause the tragacanth to cake on the sides. Mix the powdered tragacanth with the alcohol. Then add the water quickly, shaking or stirring so as to avoid lumps forming.

The following is a combination of a tragacanth mucilage with an O/W emulsion—

Powdered tragacanth	.	.	2·0 g.
Glycerin	. .	.	20·0 g.
Distilled water	.	.	72·0 ml.
Almond oil	. .	.	5·0 g.
Triethanolamine stearate	.	.	1·0 g.
			100·0 g.

METHOD. In this case the tragacanth should first be diffused in the glycerin; 65 ml. of water are stirred into this all at once. The almond oil and triethanolamine stearate are emulsified with 7 ml. of the distilled water, and finally the emulsion stirred into the mucilage.

Tragacanth is itself an emulsifying agent, and consequently the triethanolamine stearate is not absolutely necessary. An emulsion made with tragacanth alone, however, is usually coarse. A preparation of this type requires a preservative. Salicylic acid to 0·15% would meet the case. This should be dissolved in the water first by the aid of heat.

Other substances suitable for producing mucilaginous preparations are *starch, Irish moss*, and *quince seeds*. These all require heating with water—the starch to break up the granules, and the Irish moss and quince seeds to extract the mucilaginous matter.

The following formula is based on starch and glycerin—

Starch, powdered	20·0 g.
Distilled water	5·0 ml.
Glycerin	60·0 g.
Zinc oxide	15·0 g.
	100·0 g.

METHOD. Mix 4 g. of the starch with the water and glycerin. Heat with constant stirring until a translucent jelly is produced. Mix the zinc oxide with the remainder of the starch, sift, and incorporate with the jelly.

Irish moss and quince seeds being portions

of actual plants usually require washing first. They are then boiled with sufficient water to produce a thick liquid which can be strained while hot, and which sets to a jelly on cooling. Butter muslin is suitable for straining on a small scale, as it can be suspended in the form of a bag, and the marc (exhausted raw material) can then be squeezed.

Mucilages of this type, suitably diluted, are used as hair fixatives and setting lotions. They can also be combined with vanishing creams or other O/W emulsions.

Jellies and mucilages are very prone to develop growth of mould. Suitable preservatives are alcohol (5% or more), benzoic acid, sodium benzoate, derivatives of benzoic acid, which are sold under the trade names of "Nipagin" and "Nipasol," and also salicylic acid and chlorbutol.

The provisions of the Food and Drugs Act relating to preservatives do not apply to cosmetics. Consequently, preservatives can and should be used whenever there is the slightest risk of any change taking place as the result of action of bacteria or any other low form of life. Where alcohol is one of the ingredients, this is usually sufficient. Animal and vegetable oils and fats are prone to rancidity. The addition of Nipagin M. 0·15-0·25% can usually be relied on to prevent this.

CHAPTER XII

NAIL POLISHING PASTES—NAIL VARNISHES AND LACQUERS

NAIL POLISHING PASTE

THIS consists essentially of—

(*a*) Abrasives;
(*b*) Binding agents.

It needs to be just soft enough to spread easily on the nail, and should not dry up in the container. Drying is prevented in this, as in many other preparations, by the inclusion of glycerin in the formula. Glycerin has an affinity for water, and hence is said to be "hygroscopic." This is one of the reasons why it is used in vanishing creams.

The following abrasives are used—

Stannic oxide.   Pumice powder.
Kaolin.   Heavy precipitated chalk.
Zinc oxide.   Talc.

As binding materials, the following are used—

Tragacanth.   Starch.

A suitable formula for a nail polishing paste is as under—

Stannic oxide	50·0 g.
Talc	14·0 g.
Zinc oxide	7·0 g.
Tragacanth	0·2 g.
Glycerin	14·0 ml.
Distilled water	14·0 ml.
	100·0 g. (approx.)

METHOD. Mix the tragacanth with a little of the zinc oxide. Add the remainder and afterwards the talc and the stannic oxide. Sift thoroughly through a coarse sieve, say 80 mesh. Pour on the glycerin and the water, and mix till a smooth uniform paste is produced. This may appear too soft when first made, but will stiffen on keeping. The stiffness depends to some extent on the quality of the tragacanth, so the proportion may need to be varied slightly.

NAIL VARNISHES AND LACQUERS

These consist of four classes of ingredients—

(1) Non-volatile solids which leave a film on evaporation of their solution.

(2) Solvents which will dissolve the solids and evaporate fairly quickly.

(3) Volatile substances, similar to the solvents, but having higher boiling points and therefore slower rates of evaporation. These tend to

NAIL POLISHING PASTES, ETC.

remain combined with the film, and help to produce a tougher and less brittle film which is less liable to peel off.

(4) Dyes, lakes, etc.

Class 1

Celluloid and *nitro-cellulose* may be used. The former may be obtained by cleaning ordinary photographic or cinema film. When clean and dry, it requires to be cut up into small pieces. Nitro-cellulose is, however, probably the most satisfactory. It suffers from the disadvantage of being an explosive, but can be purchased damped with methylated spirit. In this condition it is safe to handle, but must be kept in an airtight container and in a dark, cool place.

Class 2

The following is a short list of suitable solvents with their boiling points (in degrees Centigrade)—

Acetone.	55°
Methyl acetate	56°
Ethyl acetate	68°
Butyl acetate	110°
Amyl acetate	125°
Ethyl lactate	135°

Class 3

The following are some plasticizers with their boiling points (in degrees centigrade)—

Ethyl benzoate	212°
Butyl oxalate	237°
Resorcinol diacetate	278°
Ethyl phthalate	295°
Butyl stearate	355°
Tricresyl phosphate	430°

A good nail varnish contains a mixture of substances covering a range of boiling points.

A formula follows, in which the reader should observe the relative proportions of the three types of ingredients, and the various boiling points and their range. With this as a basis many other formulae may be devised and tried out.

Nitro-cellulose (damped with methylated spirit)	25·0 g.
Amyl acetate	15·0 ml.
Ethyl acetate	25·0 ml.
Ethyl lactate	15·0 ml.
Acetone	25·0 ml.
Butyl stearate	1·0 ml.
Tricresyl phosphate	1·0 ml.

METHOD. Measure the solvents. Place the nitro-cellulose in a wide-mouthed bottle with the mixture of solvents. Shake or stir at intervals till dissolved. Add the plasticizers. Mix thoroughly.

It will be obvious that the larger the proportion of low boiling point solvents, the quicker

NAIL POLISHING PASTES, ETC.

will be the rate of drying. The larger the proportion of high boiling point solvents or plasticizers, the slower will be the rate of drying. The larger the proportion of nitro-cellulose the thicker the preparation.

Class 4

The colours may be in solution or suspension. For clear varnishes alcohol soluble dyes should be used. The following are typical—

Eosin. Carmoisin.
Safranin. Oil orange.

A good method is to prepare concentrated solutions of the various dyes in some suitable solvent. As the solubilities vary it may be worth while finding out what solvent is best suited to each dye. If sufficiently concentrated solutions can be prepared, the simplest way is to prepare the uncoloured base and concentrated dye solutions, of which small quantities will produce the required depth of colour. Obviously, large quantities would affect the strength of the product, by thinning down the varnish and lessening the percentage of nitrocellulose.

If any required dye is not sufficiently soluble to allow of this method being adopted, it should be dissolved in the liquid portion of the base before the addition of the nitro-cellulose. This allows of filtering out any undissolved particles of the dye stuff.

Lakes, being insoluble, can be added direct to the colourless base. About 3% is usually sufficient.

METHOD. Sift the lake. Rub it down with a small quantity of the colourless base. Gradually add the remainder.

CAUTIONS

Nitro-cellulose should be kept slightly damp with methylated spirit, as the dry product is explosive.

Some of the ingredients are highly inflammable, and so must be kept well away from a naked flame.

As some of the solvents are very volatile, evaporation takes place to an appreciable extent if the product is left uncovered.

Heat should on no account be used to facilitate solution.

CHAPTER XIII

DEPILATORIES—POWDER—CREAM—LIQUID—THALLIUM CREAM

DEPILATORIES may be divided into three classes—

(1) Preparations of chemical substances whose action is to gelatinize the keratin of the hair.

(2) Preparations of substances intended to weaken the growth of the hair, or loosen it so that it falls off or can be easily removed.

(3) Preparations of a sticky nature which grip the hair and can then be pulled off, bringing the hair with them.

CLASS 1

Depilatories in this class are usually preparations of *sulphides*. The sulphides used are—

Barium sulphide.
Calcium sulphide.
Sodium sulphide.
Strontium sulphide.

Selenides are also used in place of sulphides. Sodium selenide is said to have a similar action to sodium sulphide, but to be less irritating.

The sulphides may be prepared in the form of—
- (*a*) Powders.
- (*b*) Creams.
- (*c*) Liquids.

(a) Powders

In this case they are mixed with a suitable inert powder, such as starch, chalk, kaolin, talc or zinc oxide. Suitable proportions are—

Sulphides . . . 50%
Inert powder . . 50%

The following is an example—

Strontium sulphide . . .	50·0 g.
Starch powder . . .	30·0 g.
Talc	20·0 g.
	100·0 g.

METHOD. Mix the powders together in a thorough manner. When mixed they should be kept in a dry bottle, well stoppered. In order to make up the preparation for use, take a little of the powder and mix with sufficient water to form a paste. This should only be mixed immediately before use.

On mixing with water, the sulphides are converted into hydrogen sulphides, thus—

Calcium sulphide + Water
= Calcium hydrogen sulphide + Calcium hydroxide

The calcium (or other) hydrogen sulphide is the actual substance which gelatinizes the keratin.

(b) Creams

These may be made with a base containing starch, tragacanth, glycerin, wool fat, zinc oxide, and water. The following is an example—

Starch, powdered	8·0 g.
Glycerin	8·0 g.
Strontium sulphide	20·0 g.
Zinc oxide	16·0 g.
Wool fat	8·0 g.
Sodium sulphide	2·0 g.
Distilled water	37·0 ml.
Tragacanth	1·0 g.
	100·0 g.

METHOD. Mix the starch, glycerin and 4 ml. of the distilled water to a thin paste. Heat gently, stirring continuously, until the mass stiffens. Mix the strontium sulphide and zinc oxide together, and sift. Melt the wool fat, and incorporate the mixed powders. Dissolve the sodium sulphide in the remaining water, and work this into the wool fat and powders. Now incorporate the starch paste, and finally the tragacanth.

(c) Liquids

A suitable depilatory liquid is a solution of sodium sulphide in distilled water. Example—

Sodium sulphide	10·0 g.
Distilled water	*to* 100·0 ml.

The great drawback to all these preparations

is the smell of hydrogen sulphide, which must be disguised as far as possible. Oil of lavender may be tried, or any combination of essential oils, 1% being a suitable proportion for powders and creams. Alcohols must be avoided as they are liable to produce evil-smelling compounds known as "mercaptans."

CLASS 2

The only substances known to act in this way are the salts of thallium.

Thallium acetate has been used internally, by hypodermic injection, and in the form of a cream for application to the part affected. The first and second methods are essentially in the province of the medical profession. The third is used, although it is rather uncertain in its action.

The strength should not exceed 1%.

A non-alkaline water in oil cream is suitable as a base, the thallium acetate being dissolved in the water portion before adding to the fats. (See Water in Oil Emulsion formula, page 45.)

This type of preparation requires to be applied for a considerable period.

Thallium salts are POISONS, and therefore must be treated with respect, and the law complied with.

CLASS 3

A formula for a depilatory wax has already been given on page 78.

CHAPTER XIV

ALCOHOL—COSTING—VARIATIONS IN QUALITY

IN the formulae given, and in the text of the previous chapters, the word *alcohol* has been used without any qualification. From the chemical point of view this means *ethyl alcohol*. From the point of view of purity and strength it can be taken to mean either—

(*a*) Rectified alcohol, 95%; or
(*b*) Industrial alcohol, toilet quality, which should be also approximately 95%.

The former is preferable if the price obtained for the finished preparation is sufficient to justify the cost. The latter is generally used as the saving effected is enormous.

Unfortunately, in Great Britain permission has to be obtained to use industrial alcohol. The formulae, in which it is proposed to use it, have to be submitted to the local Officer of Customs and Excise for his approval and the product has to comply with certain requirements. The chief idea being that the product should be too unpalatable to be consumed, there should be no difficulty in obtaining permission to use it for such products as vanishing creams since no one is likely to appease his craving for alcohol by eating vanishing cream.

There is also no difficulty in obtaining permission to use industrial alcohol for making dye solutions which will subsequently be evaporated, as in dyeing powders.

It is when we come to mixtures of alcohol and water that special conditions have to be observed. Such mixtures are required to contain one or more substances, known as *denaturants*. These are substances designed to render the preparation definitely unpalatable.

Essential oils are regarded as denaturants up to a point. Thus 5% would probably be accepted as sufficient without any further additional denaturant. For solutions containing less than this proportion of essential oils, but 50% or more alcohol, the denaturant generally used is *diethyl phthalate*, 1% of this being sufficient if combined with 1% of essential oils.

Should the percentage of essential oils fall below 1%, an additional denaturant will be required, such as *quassin* to the extent of 0·035%. *Dry extract of quassia*, which is a cheaper substance, can also be used, but a larger quantity will be required, and also a brown colour is imparted to the liquid.

COSTING

Throughout this book the metric system has been employed. It is simplicity itself.

ALCOHOL 97

Apothecaries' weights and measures, on the other hand, seem designed to confuse and complicate. Nothing is what might reasonably be expected. A minim does not weigh a grain, but a little less. A fluid drachm is not the equivalent of a drachm. There are 437·5 grains in an avoirdupois ounce and 480 in an apothecaries' ounce.

A 1% solution is 1 grain in 109·47 minims.

There is one point, however, at which it becomes necessary to make contact with imperial weights and measures. That is, in pricing and possibly buying. There is, of course, no reason why one should not buy in metric quantities from a chemist or a drug house. Any such firm worth its salt will supply by the kilo or the litre, if desired. The published price lists, however, quote by the avoirdupois ounce (oz.) and the pound (lb.), and in certain cases by the pint or gallon. Also, in buying direct from manufacturers or agents it is usual to buy in these quantities.

The following equivalents are given for the purpose of calculating the quantities to be bought—

    1 avoirdupois ounce = 28·4 grammes (approx.)
    1 pound (lb.) = 453·6 ,, ,,
    1 fluid ounce = 28·4 mils ,,
    1 pint = 568·2 ,, ,,
    1 gallon = 4545·9 ,, ,,

For calculating the cost of a preparation the following method saves a good deal of trouble.

In place of grammes and mils read pounds and fluid pounds. By the latter expression is meant the volume occupied by one pound of water.

If all the quantities are expressed in grammes, and the prices of the various ingredients are given in the price list or on the invoice by the pound, the matter is straightforward. Multiply the price per pound by the number of grammes. The total of the figures thus obtained will be the cost of an equivalent in pounds to the total of the formula in grammes. Thus a percentage formula would give the cost of 100 lb.

If some of the quantities are expressed in mils, and the article is priced by the pound, multiply by the specific gravity to convert mils to grammes or fluid pounds to pounds.

If some of the quantities are expressed in mils and the price list shows the price per gallon, no correction for specific gravity is necessary. There are 10 fluid pounds to the gallon, so divide by 10 and multiply by the price per gallon. Again, there are 8 pints to the gallon, and therefore if the price list shows the price per pint, divide by 10 and multiply by 8, and then multiply by the price per pint.

If the quantity is expressed in grammes, and the article is priced by the gallon, divide the quantity by the specific gravity to convert grammes to mils, or pounds to fluid pounds.

A fluid ounce is the volume of an avoirdupois

ALCOHOL 99

ounce (oz.) of water. There are 16 avoirdupois ounces to the pound, and there are 20 fluid ounces to the pint. Therefore, a pint of water (S.G. 1) = $1\frac{1}{4}$ lb., and a pound of water = $\frac{4}{5}$ of a pint.

In some of the formulae given, the quantities of certain ingredients are expressed in grammes and of others in mils. These have been added together to give a total of 100. This procedure is only strictly correct in the case of water, of which a mil weighs a gramme. In other cases, the number of mils should have been multiplied by the S.G. to convert them into grammes before adding up. As the actual discrepancy was small, the correction was ignored. The correct way would have been to show all quantities as grammes for the purpose of arriving at the total and as mils for convenience in making up. This does not apply in cases where the preparation is made up "to" a given volume or a given weight.

VARIATIONS IN QUALITY

A factor which may give rise to serious trouble, yet which is easily overlooked or forgotten, is the variation in ingredients when purchased from different houses. That the variations occur to anything like the extent they do is deplorable, but the facts have to be faced.

Magnesium stearate, for example, will be

white and inodorous from one firm as it should be, but a yellowish white with a rancid smell from another.

Elder flower water may have a pleasant fresh smell or a heavy sickly one.

The moral is to find a good source of supply for each ingredient and then to stick to it. If it becomes necessary to change, compare the old and the new carefully. In any case, never put anything into stock without examining it. One soon gets accustomed to the physical characteristics of the various substances, the feel, colour, smell, etc. Variations in quality or mistakes in sending the wrong article may often be detected by careful examination in the light of one's experience.

INDEX

Acacia, 42
Adherence, 7
Alcohol, 24, 26, 27, 29, 51, 52, 59, 70, 73, 74, 95, 96
Aldehydes, 58
Almond oil, 36, 45, 58, 77
Ammonia, strong solution of, 53, 54, 58, 63
Ammonium hydroxide, 53
—— stearate, 53, 63
Apothecaries' weights and measures, 97
Apparatus, 10, 11, 12, 16, 23, 24, 35, 38, 44, 75
Armenian bole, 9, 32
Astringent, 29
Atoms, 6, 47, 48
Avoirdupois, 97

Bacteria, 84
Barium sulphide, 91
Benzoin, tincture of, 29
Borax, 42, 45
Boric acid, 25
Burnt sienna, 9
Buying, 97–100

Cadmium yellow, 9
Calcium carbonate, 8
—— soap, 42
—— sulphide, 91
Carmoisin, 89
Castor oil, 30, 38, 73, 74
Celluloid, 87
Chemical constitution, 4–6, 47
Chlorbutol, 84
Cleansing cream, 35
Colophony, 78
Colours, 9, 13–19, 31, 68, 71, 72, 89
Continuous phase, 41, 47, 57
Cubic centimetre, 23

Denaturants, 96
Depilatories, 91–94

Depilatory wax, 78
Diethyl phthalate, 29, 96
Disperse phase, 41, 47, 57
Drachms, 97
Dyes, 9, 68, 70, 74, 89

Emulsifying agent, 41, 42, 47, 50
Emulsions, 26, 40–46, 59
Equivalents, 97
Essential oils, 29, 30, 33, 96
Ethyl alcohol, 95
Extract of quassia, 96
Eyebrow pencils, 77
Eye shadows, 71

Filter paper, 28
Foundation creams, 61–67
Funnel, 28

Gallon, 97
Gelatin, 81
Glycerin, 24, 33, 57, 60, 61
Gramme, 1, 2, 97

Hard paraffin, 34, 35
Homogenizer, 44
Hydrocarbons, 33, 34
Hydrogen, 6, 48

Imperial measures, 97
Industrial methylated spirit, 95, 96
Irish moss, 83

Kaolin, 8
Keratin, 91
Ketones, 58
Kilogramme, 1, 2

Lakes, 9, 13, 14, 68–70, 76, 77, 90
Lamp black, 79
Lanette wax, S.X., 44
Lard, 44
Liners, 77

101

102 COSMETICS AND HOW TO MAKE THEM

Lipsticks, 72, 77
Liquid paraffin, 34, 58, 65, 66, 69
—— powder, 30–32
Litre, 23
Lubrication, 76

MAGNESIUM carbonate, light, 8, 32
—— stearate, 8
Maleic acid, 84
Mascara, 79, 80
Massing, 80
Measures, 23, 97
Mercaptans, 94
Mesh, 12, 13, 17, 21, 38
Methylated spirit, 90, 95, 96,
Metric system, 1, 96
Milligramme, 1, 2
Millilitre, mil, or ml., 23, 97
Minim, 97
Molecule, 5, 6, 47, 48
Mortars, 11
Mucilages, 81–84
Muscle oil, 30

NAIL lacquers, 86–90
—— polishing paste, 85, 86
—— varnish, 86–90
Nipagin m., 30, 84
Nipasol, 84
Nitro-cellulose, 87–90

OCHRE, 9, 18, 32
Oil orange, 9, 17, 18
Oils, essential, 29, 30, 33, 96
Opacity, 7, 62, 63
Orange lake, 9
Osmo kaolin, 8, 22
Ounce, 97
Oxygen, 6, 21, 22, 48

PARAFFINS, 33
Percentage, 2–4
Perfume, 19, 31–33, 52
Pestle, 11
Petroleum, 33
Phenol-phthalein, 49
Pigments, 9, 13, 14, 30, 32
Pint, 97

Pore cream, 38
Potassium hydroxide, 47, 48
—— stearate, 50
Pound, 97
Powder creams, 61–67
Preservatives, 84
Pricing, 97, 98
Pumice powder, 85

QUASSIN, 96
Quince seeds, 83

RECTIFIED alcohol, 95
Red lake, 9, 77
Resorcin, 38
Rice starch, 8, 21, 38
Rouges, 68–70

SAFRANIN, 89
Salicylic acid, 38, 84
Scales, 10
Selenides, 91
Sienna, 9
Sieve, 12, 38
Sifter, 11
Sifting, 13
Skin food, 36
—— tonic, 27
Slip, 7
Soap, 42, 47
Sodium hydroxide, 47, 49, 54
—— selenide, 91
—— sulphide, 91
Soft paraffin, 34, 35
Solutions, 26–30
Spatula, 11
Specific gravity, 24
Spermaceti, 45
Stannic oxide, 85
Starch, 8, 21, 38, 83
Stearic acid, 6, 47–58
Strontium sulphide, 91–93
Suet, 36
Suspensions, 26, 30, 38

TALC, 8, 21, 63, 64, 69
Tartrazine, 9, 15
Thallium acetate, 94
Thermometer, 37
Titanium dioxide, 8, 21, 62–64

INDEX

Tragacanth, 42, 66, 81-83, 85, 86
Triethanolamine, 47, 55
—— stearate, 43, 55, 59, 79
Triturations, 13-18

ULTRAMARINE, 9

VANISHING creams, 47-59
Variation, 99

WATER bath, 35
Water cosmetic, 79

Waterproof creams, 61-67
Wax masks, 79
Weights, metric, 1, 2
White wax, 35, 45, 73, 77
Witch hazel, distilled extract of, 27
Wool fat, 36, 45

ZINC oleo-stearate, 67
—— oxide, 8, 21, 38, 62, 65-67, 83
—— soap, 42
—— stearate, 8, 22, 65